My Ismail Mamouni

# Didactics of Mathematics: Part I: Learning

AF301979

**My Ismail Mamouni**

# Didactics of Mathematics:
# Part I: Learning

**ScienciaScripts**

**Imprint**

Any brand names and product names mentioned in this book are subject to trademark, brand or patent protection and are trademarks or registered trademarks of their respective holders. The use of brand names, product names, common names, trade names, product descriptions etc. even without a particular marking in this work is in no way to be construed to mean that such names may be regarded as unrestricted in respect of trademark and brand protection legislation and could thus be used by anyone.

Cover image: www.ingimage.com

This book is a translation from the original published under ISBN 978-620-6-71133-9.

Publisher:
Sciencia Scripts
is a trademark of
Dodo Books Indian Ocean Ltd. and OmniScriptum S.R.L publishing group

120 High Road, East Finchley, London, N2 9ED, United Kingdom
Str. Armeneasca 28/1, office 1, Chisinau MD-2012, Republic of Moldova, Europe
Printed at: see last page
**ISBN: 978-620-7-67800-6**

# Didactics of mathematics :
## Part I: Learning

Prof. My Ismail MAMOUNI
**CRMEF Rabat, MOROCCO**

# Table of contents

# Chapter 1 : Epistemology of learning

## Introduction

Epistemology is a branch of the philosophy of science which criticizes the scientific method, the logical forms used in science, as well as the principles, fundamental concepts, theories and results of the various sciences, in order to determine their logical origin, value and scope. On the other hand, "continental" epistemology can also deal with non-scientific objects, and the word is also sometimes used to designate this or that theory of knowledge. Jean Piaget defines epistemology as the study of the constitution of valid knowledge, which allows us to ask three major questions:

- What is knowledge (the gnoseological question)?
- How is it constituted or generated (the methodological question)?
- How do you assess its value or validity?

Epistemological study can also focus on several aspects: the ways in which knowledge is produced, the foundations of this knowledge, the dynamics of this production. This raises a number of questions: what is knowledge? How is it produced? How is it validated? What is it based on? How is knowledge organized? How does it evolve (and in particular, does it progress?) What is "good" knowledge? For a long time, epistemology focused on the "content" of science (scientific knowledge), leaving the nature of science to other disciplines, notably sociology. In recent decades, this division has become less clear-cut, under the influence on the one hand of certain currents in sociology demanding a "right of scrutiny" over this content, and on the other of certain epistemologists who deem it necessary, in order to better understand scientific knowledge, to pay attention to the concrete dimensions of the nature of science.

## History

The English label epistemology was introduced in 1856 by James Frederick Ferrier to translate the German Wissenschaftslehre (Fitche's problematic). But there are those who attribute the concept of epistemology to Eduard Zeller, who uses the German word Erkenntnistheorie ("theory of knowledge") in a Kantian sense. The word epistemology first appeared in France in 19019, in a translation of the introduction to Bertrand Russell's Essay on the Foundations of

Geometry: *"It* was only from Kant, the creator of epistemology, that the geometrical problem received its present form". This translation gives the word the meaning of "a theory of knowledge based on the critical study of the Sciences, or in a word, Criticism as Kant defined and founded it". Modern epistemology thus has its origins in the Kantian philosophy of knowledge. But it also draws on older traditions, including Cartesian ones. It was at the beginning of the 20th century that epistemology emerged as an autonomous disciplinary field.

## Rationalism (17th century)

Rationalism is an epistemological trend that holds that "all valid knowledge comes either exclusively or essentially from the use of reason". Greek philosophers such as Euclid, Pythagoras and Plato defended rationalist positions, giving primacy to ideas. More recently, the mathematicians Descartes (1596-1650) and Leibniz (1646-1716), as well as the philosopher Kant (17241804), have been associated with this trend, which favours reasoning in general, and deductive (or analytical) reasoning in particular, which moves from the abstract to the concrete as a mechanism for producing knowledge.

It's important to understand here that, for rationalists, experimentation is excluded from the mechanism for producing new knowledge. Experimentation (or interaction with reality) serves at most to verify what has been deduced, and insofar as what has been deduced is self-evident. For rationalists, the set of all possible reasonings necessarily encompasses the set of all possible experiences, and reason alone is sufficient to separate experiences that are possible in reality from those that are only possible in the imagination.

Historically, the knowledge associated with the field of geometry has played an important role in the development and justification of the rationalist epistemological position. For example, Britannica (2001) reports that Plato, in his dialogue entitled Meno, highlights the certain, universal and innate nature of knowledge by recounting how Socrates succeeded in getting a young illiterate slave to demonstrate, step by step and without teaching him, the Pythagorean theorem applied to the diagonal of a square. Later, in the early 17th century, the inventor of analytical geometry, French mathematician René Descartes, would take up the rationalist position, attempting to apply the rigor and clarity of mathematics to the realm of philosophy. ʌ' Within the rationalist trend, we can distinguish, among others, Platonism, which believes in *"an inherent harmony in nature that reflects itself in our minds"*, and Kant's criticism (17241804), which considers that knowledge depends on structures inscribed a priori in the human mind that make it possible to perceive reality. A science teacher with rationalist allegiances will obviously tend to insist on the importance of reasoning (to the detriment of experience), perhaps going so far, in extreme cases, as to completely eliminate experimentation from the student's learning process. For such a teacher, a science lesson is a sequence of analytical reasonings that the student must be able to understand, reproduce and master.

# Empiricism (18th century):

Empiricism is radically opposed to rationalism, proposing that all knowledge comes essentially from experience. We recognize this tendency in the generalizing propositions of the Greek philosopher Anaximenes (610-545 BC). The English philosophers Bacon (1561-1626), Locke (1632-1704) and Berkeley (1685-1753) are associated with this trend, which proposes that the sciences progress by accumulating observations from which laws can be extracted through inductive (or synthetic) reasoning that moves from the concrete to the abstract. For empiricists, observations are the key to understanding reality.

Deduction is excluded from the mechanism for producing new knowledge. Deduction would be no more than a temporary step in making a hypothesis, or in simplifying the description of all the observations made by scientists at a given time. Empiricists are more flexible in their definition of reasoning, especially when it comes to inductive reasoning. Indeed, since only experiments really count, the sole purpose of reasoning is to produce ideas that will enable new experiments to be carried out. Creative rather than rigorous reasoning is therefore favored, and perhaps we should call the scientific induction that takes a set of known experiments and imagines new ones abduction or conjecture. For empiricists, the lack of rigor in reasoning does not necessarily prevent it from contributing to the advancement of knowledge, since the only true rigor comes from experience, and nature is not necessarily accountable to reason.

Historically, the work of Newton (1642-1726), with its emphasis on experiments, contributed significantly to the spread of the empiricist position. The application of this method enabled Newton and his contemporaries to describe the forces in mechanics (particularly gravity) and to build a corpuscular model of light. A little later, Coulomb (1736-1806) used the same method to demonstrate the electrical force. In chemistry, Lavoisier (1743-1794) drew on Newton's work on light to lay the foundations of modern chemistry by proposing an experimental method for identifying the fundamental elements.

Within the empiricist movement, we distinguish between materialism, which proposes that everything that is not direct material experience does not exist; sensualism, which proposes that all knowledge comes from sensations; and instrumentalism, which proposes that all theory is a tool, an instrument for action, and that it teaches us nothing about the nature of reality. A science teacher of empiricist allegiance will tend to insist on the importance of experimentation by students in order to demonstrate approximate laws or test hypotheses. Reasoning to rigorously deduce these laws will be considered non-essential and, in extreme cases, may be eliminated from the student's learning process. For this teacher, a science lesson is a series of crucial experiments that the student must succeed in understanding, reproducing and mastering.

## Positivism (19th century):

Although the Greek philosopher Sextus Empiricus (160-210), who lived at the turn of the 3rd century, adopted a positivist po- sition, insisting on the suspension of all judgment, the positivist current is generally attributed to the philosopher Auguste Comte (1718-1857), as well as to the physicists Mach (1838-1916), Bridgman (1882-1961) and Bohr (1885-1962). The positivist movement is inspired by empiricism, in that it focuses on observational facts alone, but recognizes the importance of reasoning, adding that the sciences use mathematization to link experimental data$_1$ together as simply as possible.

It should be noted that positivists insist on the rigor of inductive reasoning, which enables us to move from facts to hypotheses. Positivists such as the philosopher and economist Stuart Mill (1806-1873) and the geneticist Fisher (18901962) have developed inductive methods, based on probability and statistics, to obtain probable laws from a set of measurements. However, it has to be said that to date, there is no strict inductive logic that does not contain a purely conventional component. Since inductive reasoning is indispensable (for positivists) to the evolution of science (according to Auguste Comte's famous phrase "seeing is predicting"), the theories produced have no value in themselves other than to be linked to facts. They tell us nothing about reality that is not already contained in the facts themselves. Consequently, for positivists, "science describes the how of things, without being able to say anything about their why".

Historically, this clear distinction between observations (the how) and mathematical models (the why) is particularly important in understanding what led positivists to distinguish themselves from empiricists. For example, the experimental work of Dalton (1766-1844), which founded chemical atomism, raised the fundamental question of whether atoms really existed. The empiricists of the time generally believed that atoms, since they were necessary to explain experimental results, really did exist. Positivists fiercely opposed the existence of atoms because they were not directly observable: atoms were models (the why) to explain experiments (the how). For positivists, models were human creations with strictly no value other than being useful. Positivists thus categorically opposed anything in scientific models that was not directly observable. For example, according to Feigl (2001), for positivists, the infinitesimals used by Newton in calculating the motion of bodies undergoing forces were mere mathematical artifices; the vacuum between atoms could not exist, a true medium was preferred: the ether; Newton's notion of absolute space and time could not be real, space and time always had to be measured in relation to something material. For positivists, the fact that models have no value in themselves opens the door to the possibility that several different (and even contradictory) models can account, with equal effectiveness, for the same observations. In this way, the emergence of the positivist movement has, in a way con, encouraged the multiplication of models. The debate between Newton's corpuscular model of light and the undulatory model proposed by Fresnel (1788-1827) is a case in point. Positivism is often associated with an excessive tendency towards classification and organization. Positivists tend to believe, for example, that there is a universal experimental method with precise steps that guarantees the progression of science. Positivism is also associated with the subordination of the sciences to one another according to a strict classification, and with a universal order of knowledge and human society. In extreme cases, the distorted theses of positivism have given rise to scientistic ideology.

According to Kremer-Marietti (1993, pp. 10-11), within the positivist trend, we can distinguish between the conventionalism of Poincaré (1854-1912), who proposes that hypotheses have no cognitive value in themselves, the pragmatism of James (1842-1910), who proposes, according to Le Moigne (1995, p.. 55), that "the true consists simply in what is advantageous for thought", and the logical positivism of Carnap (1891-1970), which proposes that the cognitive processes involved in the elaboration of representations must be capable of being constructed or reconstructed. Logical positivism is sometimes presented as one of the precursors of constructivism.

A positivist science teacher will tend to recognize the complementary importance of experimentation and reasoning in student learning, emphasizing the process of statistically analyzing a set of measurements to obtain the simplest possible model. For this teacher, a science lesson is a series of experiments that students must be able to understand, reproduce, master and logically link together through rigorous inductive reasoning.

# Constructivism (20th century) :

Some of the Greek sophists' ideas can be associated with the heritage of the constructivist position. For example, Heraclitus' (550-480 BC) conception of the ambiguity of reality and Protagoras' (485-410 BC) formula: "Man is the measure of all things". The constructivist movement emerged in the 20th century thanks to the Dutch mathematician Brouwer (1881-1966), who used the term constructivist to characterize his position on the question of foundations in mathematics, in opposition to Hilbert's formalist position. Mathematicians who have attempted to answer the question of foundations in mathematics are usually grouped into three schools: the logistic school, which attempts to base the whole of mathematics on the logic of propositions; the formalist school, which attempts to demonstrate the consistency of all the fundamental axioms of mathematics; and the constructivist school, which accepts as true only what can be constructed (in a finite number of steps) from ideas that intuition accepts as true.

The first two schools encountered insurmountable obstacles. Indeed, members of the logistic school found it impossible to completely define the logic of mathematical construction without using results from mathematics. For the formalists, Godel demonstrated in 1931 that any theory powerful enough to encompass integer theory could not be shown to be consistent. In the end, even the constructivist school had^ to come to terms with the fact that it could not cover the entire field of classical mathematics with a single set. It therefore seems impossible to unite mathematics into a coherent, complete system that does not contain a subjective component that constructivists call intuition. "Mathematics walks on two feet, intuition and logic, [...] the first step is intuition; logic comes next...".

The constructivist position was taken up by the Swiss psychologist Piaget and Garcia, who questioned the possibility of always obtaining objective relationships on which to base science. The absence of objective relationships obviously invalidates any formal verification process. By renouncing objectivity, the constructivist movement proposes that the sciences construct (rather than reveal) a possible reality based on successive cognitive experiences. Constructivists do not reject the existence of an ultimate reality, but they assert that it cannot be known. The constructivist movement, which is not very present in traditional scientific circles, plays an important role in psychology and didactics. In psychology, for example, we use the term constructivism to describe the model adopted to apprehend a subject's cognitive activity, while in didactics we use the term to describe certain teaching procedures where the student is at the heart of learning.

Within the constructivist movement, we distinguish between trivial constructivism, which proposes that "knowledge cannot be passively transmitted, but must be actively constructed by the subject", and radical constructivism, which takes up the previous proposition, adding that "cognition must be seen as an adaptive function that serves to organize the world of experience rather than to discover an ontological reality". A science teacher with constructivist allegiances will tend to insist on the arbitrary character of cognition.

or subjective nature of scientific models, encouraging students to construct their own knowledge. For this teacher, experimentation only serves to verify the internal coherence of the construction. For this teacher, a science lesson corresponds to a series of models currently recognized by the scientific community, which the student must succeed in understanding, constructing and mastering. A teacher who adopts a constructivist conception of science will obviously also tend to adopt constructivist teaching procedures where the student is at the heart of learning.

# Realism (20th century):

The Greek philosopher Aristotle (384-322 B.C.), in his concern to build some of his models from systematic observations of nature, defended a position that can be described as realist. Realism proposes that scientific models are approximations of an objective reality that exists independently of the observer. In contrast to rationalism, empiricism and positivism, this current does not retain a precise mechanism for the progression of knowledge, but rather recognizes the complementarity of different approaches. The physicists Planck (1858-1947) and Einstein (1879-1955) are generally associated with this movement.

It is the recognition of the existence of a reality towards which scientific models tend that distinguishes realism from constructivism. In contrast to the radical constructivist proposition that the observer constructs reality, realism proposes that the observer is part of reality. In other words, reality reacts[1] con coherently (insofar as' reality is coherent) whatever model is chosen to describe it.

Historically, the work of Michelson (1852-1931) on the speed of light and that of Einstein on relativity in 1905 helped to diminish the influence of the positivist position (in favor of the realist position) by seriously calling into question the necessity of the notion of ether hitherto defended by the positivists. [1]Similarly, the work of Rutherford (1871 - 1937) on the atomic nucleus and that of Bohr (1885-1962) on electron orbits around the nucleus reinforced the hypothesis of the real existence of atoms, which positivists had opposed from the outset. In this context, the realist position distinguished itself from the positivist position by recognizing a certain reality in the models developed, which were intended to be increasingly accurate approximations of a single reality.

Realism is very present among contemporary scientists, who distinguish between naive realism, associated with *"the tendency to take the model as reality"*, and critical realism, which proposes that *"scientific theories are successive approximations of reality"*.

A realist science teacher will tend to emphasize the complementary roles of inductive reasoning, deductive reasoning and experimentation in the search for new scientific knowledge, to insist on the difference between models (which are produced by scientists) and reality (which exists independently of models), and to recognize a subjective and creative component in the development of scientific theories. For this teacher, a science lesson corresponds to a series of experiments, reasoning and models that students need to understand, construct and master in order to predict the world around them.

# Chapter 2 : Learning Planning

## General

Didactics in general, and mathematics in particular, studies the interactions that can take place during a teaching or learning situation between an identified piece of knowledge, a teacher dispensing this knowledge and a learner receiving this knowledge. No longer content to treat the subject to be taught according to pre-established schemes, it posits as a necessary condition the teacher's epistemological reflection on the nature of the knowledge he or she will have to teach, and the consideration of the learner's representations (the learner's epistemology) in relation to this knowledge,

## Didactic Triangle

It is the schematic representation of the didactic system that appears in any transfer of knowledge between a teacher and a taught, and is shaped by the interactions produced between the following poles: Knowledge, Teacher, Taught.

The main purpose of this representation is to contrast with classic linear teacher-student schemes. It is an attempt to understand and model a more complex situation.

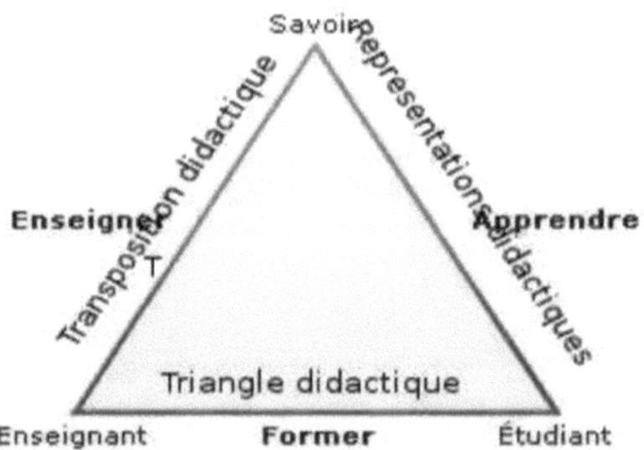


## Didactic Contract

A *"well-functioning" classroom is* determined by the right distribution-

explicit or implicit-of responsibilities and expectations between teacher and students. Teachers and students each expect something of the other. This "something" relates to teaching and learning. The effectiveness of the relationship depends on mutual understanding of each other's intentions, which determines the respective roles of student and teacher in the classroom in relation to knowledge. The didactic contract is the set of reciprocal obligations that each partner imposes or believes they impose, explicitly or implicitly, on the other in relation to the knowledge being taught. The didactic contract is the result of implicit negotiation, and defines the didactic situation (teaching-learning conditions).

## Didactic approaches

Even if the object of mathematics didactics is the same: to study, during a teaching-learning situation, the interactions produced between the three poles of the didactic triangle: Knowledge, Teacher, Taught. A number of approaches and methodologies can be distinguished, most notably the following

- Epistemological, defined as a reflection on teaching knowledge. It is concerned with their cognitive nature (knowledge or know-how); their epistemological status (scholarly knowledge or social knowledge); the methodology of their construction (transposition or elaboration of knowledge); their institutional history ...

- Psychological, defined as research into the conditions of knowledge appropriation. It is less concerned with the concepts and notions themselves, than with the conditions of their construction: the prerequisites they presuppose, the ordinary representations learners have of them, the different kinds of obstacles they may give rise to...

- Sociological, defined by research into didactic intervention. It is primarily concerned with the organization of teaching situations, the construction of didactic cycles or sequences, adaptation to the type of audience, in short, with the approach to the classroom and its own functioning.

## Didactics vs. Pedagogy

Opposing pedagogy and didactics is absurd. These two fields are obviously complementary, and it's in the practitioner's best interest to take an interest in the results published by these two branches of research if he wants to increase the effectiveness of his teaching.

Pedagogy is defined as any activity undertaken by a person to develop specific learning in others. To clarify the meaning of the term pedagogy, it is important to differentiate between pedagogy and didactics.

- The pedagogue seeks to answer questions of direct relevance to his or her educational work: *"What do we know about human learning that will enable us to build effective teaching strategies?"* or "What would be the most effective teaching method for a given type of learning?" or "How can we promote learning to read using a small newspaper in the primary classroom?

- The pedagogue thus appears as a practitioner who is primarily concerned with the effectiveness of his action. He's a man of the field, and as such is constantly solving concrete teaching-learning problems. The main source of his "pedagogical intuition" remains action and experimentation, from which he draws validation and encouragement;

- The didactician, on the other hand, is first and foremost a specialist in the teaching of his discipline. He is primarily concerned with the notions, concepts and principles that are to be transformed into teaching content. He also assesses the level of his pupils (individual difficulties, personal representations, etc.) to identify the epistemological or psychological obstacles he needs to overcome to "get them to learn". The job of the didactician is therefore essentially one of information processing: identifying and transforming "scholarly knowledge" into "knowledge to be taught". This is known as "didactic transposition".

## Transposition Didactique

We owe the concept of "transposition didactique" to Yves Chevallard (1985), who, noting the arrival of new knowledge in the teaching system, asked himself the following two questions:
1. Where do these new teaching objects come from?
2. How did they get there?

Chevallard (1985) defines didactic transposition as "a knowledge content that has been designated as knowledge to be taught, which then undergoes a set of adaptive transformations that make it suitable for use as a teaching object. The work that transforms an object of knowledge into an object of teaching is called didactic transposition. The process of didactic transposition can be summarized as follows:

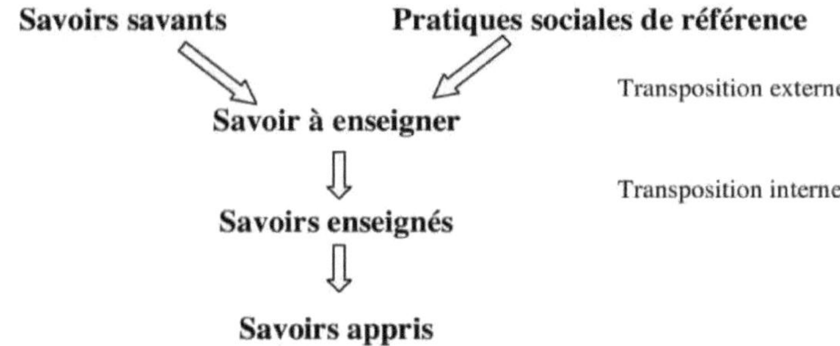


- By **"scholarly knowledge"** we mean knowledge that is recognized as relevant and validated by the specialized scientific community, which legitimizes this knowledge and confers on it a label of accuracy and interest (Le Pellec, 1991; Audigier, 1988).

- The **"knowledge to be taught"** is that "which is described and specified, in all "official" texts^ (programs, official instructions, commentaries...); these texts define content, standards and methods" (Audigier, 1988).

- **Taught knowledge"** is that which the teacher has constructed and will apply in the classroom. It's what's taught during class time.

- **Learned knowledge"** refers to all the knowledge acquired by a given learners.

Didacticians distinguish two stages in didactic transposition:

The first, called **"external didactic transposition", takes** place outside the teaching system, outside the classroom. It is regulated by what Chevallard (1985) calls the "noosphere", literally "the sphere where we think". The noosphere is therefore the set of people who think about teaching content: academics interested in teaching issues, representatives of the teaching system (e.g., the president of a teachers' association), textbook authors, school inspectors, representatives of society (e.g., the president of a parents' association) and representatives of the political world (the Minister of Education, his or her department head(s), etc.).

The second stage is called **"internal didactic transposition".** It consists in adapting and transforming the knowledge to be taught, as it appears in curricula and textbooks, and consequently the scholarly knowledge from which it derives, into taught knowledge. It is the work of teachers and their classroom practices. As a result, taught knowledge is difficult to identify (Le Pellec, 1991). It's normal to assume that taught knowledge is necessarily different from scholarly knowledge, since it has neither the same origin, nor the same function, nor the same destination. It would be incongruous for a researcher, after presenting the results of his research, to propose homework to his listeners. The researcher's function is to research, to find if possible, not to teach. The teacher, on the other hand, will have to devise educational activities, set up exercises and produce supporting documents. The teacher's job is to increase the likelihood of students appropriating knowledge. The distance separating scholarly knowledge from taught knowledge can therefore be considerable, as illustrated by the following scenario (Clerc, Minder, Roduit ;2006):

- A researcher communicates the results of his research to his peers by publishing an article in a scientific journal;
- A specialist journalist writes a popular article about it;
- The editor of a manual refers to the previous publication;
- A teacher uses the manual as inspiration for a teaching sequence on the subject of the article.

Tardy (1993) suggests some examples of the transformations that taught knowledge must make to scholarly knowledge: Technical terms, generally reserved for specialists, should be avoided in favor of everyday words. Terminological transposition means talking about the same thing in a different way. Knowledge taught must (often) confine itself to presenting the results of research as truths, or as real and true facts. In this way, taught knowledge is not open to discussion, whereas scholarly knowledge is often the subject of heated debate. Taught knowledge makes far more frequent use of examples to illustrate its points than scholarly knowledge.

It should be noted that there are some courses of study in which the reference knowledge is not the only scholarly knowledge, either because such knowledge simply does not exist (as in the case of physical education, for example), or because the purpose of the course of study leads it to favor another reference. For example, for a heating technician, the notion of energy dissipated during the path of a wave constitutes a loss of efficiency, whereas for a tele-communication technician, it's a gain, since this energy enables the heating he wants to produce in the material. Thus, social reference practices refer to all the social activities (lived, known or imagined) that will serve as a reference for constructing the knowledge to be taught and the knowledge to be taught. They enable students to give meaning to what they learn, and teachers to give meaning to what they teach. It comes down to asking the question: "What's the point in society?" (Martinand, 1985).

# Chapter 3 : Learning management Didactic Contract

## Introduction:

According to Brousseau (1990), the didactic contract is the set of teacher behaviors expected of the student, and the set of student behaviors expected of the teacher. This contract is the set of rules that determine explicitly, to a small extent, but above all implicitly, what each partner in the didactic relationship will have to manage, and for which they will be, in one way or another, accountable to the other.

The rules of the game (Brousseau; 1990) in all didactic situations, is that the teacher tries to let the pupil know what he wants him to do, but cannot say it in such a way that the pupil simply has to carry out a sequence of orders. In this way, a didactic contract is negotiated that will determine explicitly, but above all implicitly, what each partner will be responsible for managing. In this way, when a contract is established, each partner knows the other's expectations, without having to state his or her own.

However, the didactic contract presents the teacher with a veritable paradoxical injunction: everything he does to "push" the pupil to behave as he expects, sometimes tends to deprive the latter (the pupil) of the conditions necessary for understanding and learning the notion in question: if the teacher says what he wants, he can no longer get it.

## Implicit character :

One of the major problems of the didactic contract is its implicit nature. Everything happens in the school situation, as if the partners had to respect clauses that have never been discussed and are only made explicit when they are broken. This is not because the teacher is trying to hide something from his pupils, but because both he and his pupils are bound by this contract, which goes beyond them and characterizes the teaching situation. The famous example of the captain's age illustrates this implicit nature: A teacher asks his pupils the following question: "There are 26 sheep and 140 goats on a boat. How old is the captain? Surprisingly, out of a sample of 97 students, 76 gave the captain's age using the numbers in the statement. Stella BARUK (2016) comments on this paradoxical situation with the following significant words: Teaching mathematics as it is currently done turns students into "automaths", since they can answer absurd questions in an absurd way.

**Two ways of seeing.** To understand this paradox, we need to imagine two points of view:

- Student's point of view: Since the teacher has given statements, you must use all the data to answer the questions asked;

- Teacher's point of view: not all the data in the statements need to be used. The student must sort out what is relevant to solving the problem at hand.

## Implicit system of reciprocal obligations.

The didactic contract, largely implicit, determines what each partner (teacher and taught) is responsible for managing, and for which they will be accountable in one way or another to the other. The didactic contract defines the student's job, as much as the teacher's job, neither of which can replace the other without collapsing the learning task. The didactic contract depends first and foremost on the teaching strategy adopted. Pedagogical choices, the style of work required of students, training objectives, the teacher's epistemology and assessment conditions are all essential determinants of the didactic contract, which must be adapted to these contexts.

## Constraints and dependencies. Here are two examples:

- *Static aspect:* the didactic contract always pre-exists and over-determines the didactic situation. The teacher is bound by it just as much as the student, insofar as he or she is concerned. The contract is never static: it can evolve in the course of the teaching activity. The acquisition of knowledge by pupils is the fundamental challenge of the didactic contract. At each new stage, the contract is renewed and renegotiated. Most of the time, this renegotiation goes unnoticed.

- **Avoiding the worst**: the didactic contract is most apparent when it is transgressed by one of the partners in the didactic relationship. A large proportion of students' difficulties can be explained by the effects of a poorly defined or misunderstood contract (the student doesn't know exactly what is expected of him or her). Many misunderstandings and feelings of being bullied stem from an ill-adapted or misunderstood didactic contract. Pupils' desire to adapt can come up against the fickleness of a teacher whose wishes are never clear. Such situations can lead to school refusal and, in extreme cases, failure.

**Ideal scenario.** A teacher who wants to ensure the success of his didactic contract with his pupils is expected to respect the following attitudes:

- **A reminder of the rules:** during a session designed to teach a pupil specific knowledge, the pupil interprets the situation presented to him/her, the questions put to him/her, the information provided and the constraints imposed on him/her according to what the teacher reproduces, consciously or unconsciously, on a repetitive basis. In a classroom activity, everything functions as if the partners (students and teacher) had to respect clauses that have never been stated, let alone discussed.

- **Reminder of the objective: This is** not a real contract, since it is implicit and non-negotiated. However, the teacher's intentions must be clear to the students. When implementing classroom activities, the teacher's attitude is important. Once the project has been determined, the first thing to do is to see it through to the end, to take stock of what we know, to determine our needs, to start researching and writing.

- **Favoring dialogue:** during moments of confrontation, everyone's voice is listened to, and integrated into the collective project, as a result of the habit of giving advice; the individual is never judged, evaluated or graded. Group discussions are free, unencumbered by adult intervention, and progress, questioning and pauses in front of obstacles never give rise to judgment, evaluation, assessment or appraisal...

- **How to organize:** When researching, students know that the teacher has the answer to their question, or at least an answer. What if all they have to do is ask the teacher a question? Answering a question with a question, organizing the group's ideas around the question, bouncing back on questions, helping to list information-seeking strategies, refusing the obvious... are all necessary conditions for students to engage in a research process.

- **A few "tips":**

  ○ Grades and rankings are always a mistake;

  ○ Talk as little as possible;

  ○ The child does not like herd work to which the individual must conform. He likes individual work or teamwork within a cooperative community;

  ○ Order and discipline are necessary in the classroom;

  ○ Punishment is always a mistake. They are humiliating for everyone and never achieve the desired goal;

  ○ The new life of the school presupposes school cooperation, i.e. the management of school life and work by users, including the educator.

- **Attitude in the classroom.** The teacher's attitudes must be made clear to the students, so that they can understand their purpose. "I know the answer to your question, but what I'm doing here with you is guiding you to learn, to move forward with the project. Let's search together. How are we going to do it*?"*

**- What problem does a didactic contract solve?**

  ○ How do you disseminate this idea without reducing its novelty for others? It's not enough for the inventor to designate it, others must expect it; but how can they expect an idea they don't have?

  ○ This is the problem of every teacher: to create for others the conditions that will make what he's teaching them necessary, and at the same time to conceal its novelty by presenting it as a new form of the already known. In other words, posing the problem that the new idea solves and presenting the solution as an organization of known elements, long before showing the importance of the problem and the originality of the construction;

- ○ The teacher is led to propose an activity that makes sense to the student independently of the teaching content, because he or she is committed to teaching even before the student has begun studying.

# Failed didactics

## Example by Chevallard (1985):

In 4th grade, the student who, in response to the question: Factorize "4x2-36x", would answer:

$$4x^2 - 36x \ = \ 4x^2 - 2.2.x.9 + 9^2 - 9^2 = \ (2x - 9)^2 - 9^2 = \ (2x - 9 + 9)(2x - 9 + 9) = 2x(2x - 18)$$

In this way, the student would have demonstrated an unusual ability (for a 4th grader) to recognize algebraic formulas, but an inability to recognize the type of problem-situation he was faced with. He would have applied his knowledge of remarkable products, when asked to do a simple factoring. His response behavior, however valid in principle, would nevertheless be "irrelevant to the didactic contract patiently woven by the teacher".

## Example of multiplication

Teacher presentation of multiplication: The multiplication of an integer "a" by the integer "b" is an integer "c" that expresses the sum of "b" integers equal to "a", otherwise: ab = a+a+a+a+a ... +a; where a appears b times. For the student: multiplication is repeated addition (an object already known). Multiplication is only a problem in the teacher's discourse (novelty). Justification of new knowledge real knowledge: once named, we can talk about it and ask questions about it. The pupil can "multiply": a gesture that his teacher or parents will recognize as being part of this operation. Use of knowledge and contract : The teacher: "How much is four times three? The pupil "four times three makes twelve".

Interpretation: the lesson is understood, we can move on to the next one. What knowledge was used in 4+4+4? Answer: repetition of addition (not multiplication!). Student question: When should multiplication be used?

The teacher's response (the paradoxes of the didactic situation) :

- If the teacher shows the student what to do, the student carries out an order but does not apply his or her own knowledge;

- Similarly, if the teacher proposes a mathematical situation whose solution has been taught to the student, the student reproduces and quotes a piece of knowledge.

The teacher must therefore put the student in the situation of having to solve a mathematical problem for which he or she has not been taught the solution, and assume responsibility for any failure (no professional would accept such a contract).

But no student can reproduce in the time allotted knowledge that has taken centuries to acquire.

What the teacher and student expect of each other cannot be agreed in advance. Thus, the student expects the teacher to teach sufficient knowledge. The teacher can only show what is indispensable, and expects the student to produce knowledge on his own. The teacher must therefore play a double game with the pupil: giving and retaining at the same time.

**Negative effects.** According to Brousseau, the acquisition of knowledge by pupils is the fundamental challenge of the didactic contract. At each new stage, the contract is renewed and renegotiated. Most of the time, this renegotiation goes unnoticed. This type of difficulty is encountered in the transition from understanding the lesson (assimilating) to solving the exercise (applying), where there is a break in what the teacher expects the student to do, which is not always explained. Ongoing negotiation of the didactic contract tends to lead to a downward revision of learning objectives. The effort required of students may appear too great. Teachers want their students to succeed. They tend to make the task easier for them in a variety of ways: abundant explanations, teaching "little tricks" for succeeding with problems. These attitudes are real breaches (on the part of the teacher) of the didactic contract, the main aim of which is to get students to master the knowledge that is being avoided.

**TOPAZE effect.** When a student encounters a difficulty, the topaz effect consists, in one way or another, in overcoming it for him or her. When the help is decisive, the pupil does not make the necessary effort himself, which would bring him to a level of understanding that would enable him to achieve the intended learning. The primary objective is therefore not achieved. The topaz effect is very common, and is often necessary to unblock students in difficulty. The teacher must be aware of how it works, and of its consequences.

Based on the play by Marcel Pagnol: Topaze dicte en se promant.

- "Sheep... sheep... were t-en safe... in a park; in a park.

- He leans over the student's shoulder and starts again). Sheep... sheep... (The student looks at him, bewildered.)

- Come on, my child, make an effort. I say moutonse. Étaient (he starts again with finesse) étai-eunnt. That is, there wasn't just one sheep. There were several moutonsse."

For the student, it's primarily a problem of spelling and grammar. Faced with repeated failures, Topaze negotiates 'downwards the conditions under which the pupil will eventually put an s The teacher ≪ suggests 'the right answer, concealing it under increasingly transparent didactic coding The teacher takes 'over most of the work the targeted knowledge disappears completely

Example of the transition from multiplication to addition:

- The teacher: 5x4

- The student: 4+4+4+4+4= 8+4+4+4=12+4+4=16+4=20

The teacher simplifies the task by ensuring that the student obtains the right answer by simply reading the teacher's questions, rather than by carrying out a genuine mathematical activity specific to the proposed structure.

**The JOURDAIN effect.** A student's trivial behavior is interpreted as the manifestation of learned knowledge. This avoids the need to learn this supposedly acquired knowledge, and each partner in this perverted didactic relationship is satisfied to get away with it. But there is a breach of contract on the part of the teacher.

- Example of Le Bourgeois Gentilhomme

◦ Philo: So you're going to write prose.

◦ M. Jourdain: No, I don't want prose or verse. Philo: It has to be one or the other.

◦ Mr Jourdain: Why?

◦ Philo: Because prose or verse is the only way to express yourself.

◦ M. Jourdain: Is it just prose or verse?

◦ Philo: Yes, sir. Everything that isn't prose is verse, and everything that isn't verse is prose.

◦ M. Jourdain: And when we speak, what is that? Philo: Prose!

◦ M. Jourdain: When I say "Nicole, bring me my slippers and my nightcap", is that prose? Philo: Yes, sir!

◦ M. Jourdain: By my faith, I've been saying prose for over forty years without knowing anything about it. Philo: That's what it is to be educated, sir.

the philosophy teacher reveals to Jourdain what prose and vowels are The teacher recognizes the clue of a scholarly knowledge in the students' answers avoid the statement of failure the student deals with an example, and the teacher sees the structure insert knowledge into familiar activities

- An example from mathematics

◦ Student: 2x1=2; 1x2=2

◦ Professor: That's good, you know that 1 is neutral for multiplication and multiplication is commutative.

- ○ The pupil obtains the correct answer by means of a banal acknowledgement, and the teacher attests to the value of this activity by means of a learned mathematical and epistemological discourse.

**Effect of METACOGNITIVE SLIDING.** Taking a technique, supposedly useful for solving a problem, as the object of study and losing sight of the real knowledge to be developed.

- Functioning Replace a problem for which the mathematical knowledge to be taught provides the solution with a problem for which the material solution can be easily obtained. Interpret this success as sufficient proof of the construction of the targeted knowledge.

- Example Group structure: students are asked to swap yoghurt pots exhaustively, after which it is explained that they have studied "a mathematical structure of a finite group".

**Effect of INCOMPREHENSIVE WAITING.** Believing that a response expected from students is self-evident.

Example:

- question posed by a middle-school history teacher: "In the Middle Ages, townspeople raised ... ? "

- Student responses: "pigs, children, ... . "

- Expected response: "Cathedrals! "

## Effect of ¡'Analogy' Abuse

**Function** replace the mathematical construction with an explanation based on the manipulation of substitute symbols whose analogical use requires new explanations, etc. The use of analogical notations was supposed to produce the same knowledge as that of ordinary mathematical notations

**For example,** drawing arrows in both directions between the first names of members of the same family to explain the "equivalence relation"; for the students, the meaning of this activity is not that it is the analogue of a mathematical activity they have no idea about.

# Didactic representations

## Definition:

Didactic representations are generally considered to be systems of knowledge that a subject spontaneously mobilizes when faced with a question or problem, whether or not it has been learned (Reuter et al., 2007]). They refer to particular ways of reasoning, based on an explanatory model pre-existing formal learning. Representations form a system: a system of ideas and explanations that constitute the students' frame of reference. Representation is a process: it is evolving and personal. It enables the subject to add to the system what he or she encounters and integrates in the course of his or her experience, whether private or academic (Halté, 1992). It's "already there", the fruit of initial experience.

For André Giordan (1996), representation (or conception) is made up of various elements that interact completely:

- The problem: representation refers to the set of questions that induce or provoke its implementation;

- The frame of reference: the representation is based on a set of other representations that form a system and are mobilized by the subject to produce his new representation;
- Mental operations: representation is the product of invariant reasoning enabling the subject to relate elements and make inferences;
- The semantic network: this interactive organization produces a network of meanings capable of giving the representation a very specific meaning;
- Signifiers: all signs and symbols referring to the subject's way of expressing himself. André Giordan and Gérard de Vecchi have proposed replacing the term "representation" with "conception" (Giordan, de Vecchi, 1987).

## Advantages-Disadvantages :

The value of the conceptions constructed by students is that they provide them with a grid for reading and predicting the world (Giordan, 1996). These interpretive grids enable them to solve given problems by implementing "cognitive strategies" (Halté, 1992). In effect, they testify to an activity of mental construction of the real, the least advantages of which are their functionality and operationality. In fact, they are very practical, in that they are even capable of explaining certain things we sometimes have no experience of! (Develay, 1992).

Giordan and de Vecchi also point out that conceptions enable significant cognitive savings to be made, highlighting the fact that while it is costly to transform one's own explanatory models, it is more comfortable to use tried-and-tested schemes (Reuter et al., 2007). The downside is that learners are locked into rigid frameworks that prevent them from making any cognitive progress. Since they can only grasp the world through these frames, conceptions become the student's "intellectual prison" (Giordan, 1996).

## Origins :

Didacticians (Astolfi, Develay, Brousseau, ... ): attribute at least five origins to conceptions:

- **Psychogenetic (Piaget)**: these conceptions are due to the incomplete development of the child. Adherence to the child's intellectual functions (dualism, anthropomorphism, animism; egocentrism, artificialism, realism) hinders consideration of objective reality;

- **Epistemological (Bachelard)**: there are ways of thinking that generate obstacles, such as opinion and the "impure complex of first intuitions". An example of an epistemological obstacle: understanding that there are infinite numbers between 13 and 14. Other researchers postulate that the obstacles encountered by students relate to the very nature of knowledge;

- **Didactics**: the difficulties here are generated by the didactic situations themselves, the way in which school knowledge constructs a reality capable of instituting conventions that are no longer called into question. An example of a didactic obstacle is the way in which the planisphere is presented, with equivalences between north, up and over;

- **Sociological (Moscovici)**: in this case, they stem from social representations and prejudices. For example, common thinking on the necessarily exogenous reason for illness prevents us from thinking about genetic diseases;

- **Psychoanalytic (Freud)**: concepts are based on fantasy, psychic content, affect and the individual's personal history.

## Concept and Design: Continuity or Discontinuity

The question often asked by didacticians is: is the transition between conception and target concept to be seen as a break or as continuity? There are two types of answer-approach:

- **The constructivist approach** postulates that learning is not reduced to a process of linear, vertical transmission of knowledge, but is the product of the transformation of conceptions through the aggregation of new knowledge within the subject. All successful learning is understood as a change in conceptions (Giordan). An ideal vision would quickly emerge, in which this change is seen as a smooth path from the student's conception to the scientific concept. But is this path linear and continuous, or is it marked by wanderings and ruptures? For Jean Migne, representation and concept refer to two distinct modes of knowledge, one figurative, the other operative.

There is therefore no difference in degree between them that would place them on the same continuum. Even so, for Astolfi, they share relatively important common features. They can be stated, they are operative within a given domain of validity, and they are predictive in nature;

- **The "conceptual change" approach**: (Joshua and Dupin), propose to move a conception from a rudimentary state to a more evolved one, by moving it up degrees towards greater abstraction. According to Postner, this might involve presenting students with knowledge to be acquired that would be more satisfying and productive than their own representations, since it would be more plausible and intelligible. According to Driver, surprising experiments or counter-examples could also be proposed, with the aim of destabilizing their convictions and provoking a cognitive conflict, before presenting them with scientific models.

Other authors advocate another approach, that of setting up scientific debates in the classroom. First, a paradoxical initial situation is posed, with the aim of leading students to formulate hypotheses, which are no more and no less than the expression of conceptions that need to be changed. The verbalization of these hypotheses, or spontaneous explanatory schemes, gives them the status of "first models". During the ensuing debate between peers, hypotheses are eliminated and selected. New knowledge contributions are then proposed, requiring explicit validation, and given the status of "test experiments". This socio-constructivist approach is based on the assumption that the area of freedom thus created is sufficient to enable social inter-action conducive to learning.

Only in the case of epistemological obstacles can we speak of a rupture, i.e., when the student's explanatory model and the scientific explanatory model do not belong to the same register of explanation, or frame of reference. This is particularly the case when the targeted knowledge is derived from experience, observation or common sense, since there can be no continuity between common sense, that "impure complex", and scientific knowledge (Joshua and Dupin).

The problem remains that the conception cannot be easily overcome, either by a simple awareness on the part of the students, or by a contradictory presentation by the teacher, or by an isolated counter-example, or by confrontation with peers alone. And when this works, it's because it wasn't a genuine epistemological obstacle. As these can neither be denied nor destroyed, they must be overcome (Joshua and Dupin).

## Design dimensions

There are two dimensions

- **The structural dimension**: Didacticians speak of the existence of a stable structure responsible for the surface manifestations of conceptions. The hypothesis is that this structure is the product of mental processes pre-existing intellectual activity, understood as invariants of thought operating at the heart of the "black box". André Petitjean has thus distinguished Representation, as a socio-cognitive activity "through which each individual categorizes and interprets objects in the world" from representations grasped as the products of ordinary thought in its multiple materializations (beliefs, discourses and behaviors);

- **The functional dimension**: This introduces the idea that design is above all a particular manifestation, inscribed in a given situation, at a given moment, and that these variables must be taken into account if we are to grasp the design at work in learners' expression (Astolfi). We need to bear in mind that what we perceive in students' conceptions are always responses to a teacher's prompting in a particular production context, to which these responses must be related. Moreover, every conception is relative in nature, since it organizes knowledge in relation to a particular problem (Migne), recognized as belonging to a given field of validity, and thus providing an original and appropriate response. It should also be remembered that a conception never operates in isolation (Giordan and de Vecchi), but is part of one or more systems of explanation.

Astolfi and Develay also point out that, in their response, students seek to situate themselves in relation to various reference points, such as the teacher's supposed expectations, the self-image they wish to project and the hoped-for adjustment to the group's point of view. These are indeed "circumstantial strategies", as Astolfi puts it. Immediately available to the student and the result of didactic custom, they have a strong functional and operative value.

Added to this is the fact that the observer's interpretation is as decisive as it is relative, since the learner's expression is decoded through the observer's conceptual frameworks. Astolfi and Develay, calling for caution, provide astonishing examples showing the gap between the student's statement, a marker of conception, and the interpretation made of it.

# Didactic obstacle

### Obstacle types:

A child's comment: "When you go to the South Pole, do you feel like you're upside down?" The question here is, how do we deal with the obstacle? Ignore it without ignoring it? Avoid it and get around it by posing the problem differently? Or do we follow a strategy of "dealing with it in order to deal with it"? There are three kinds of obstacle: epistemological: nature is difficult to understand (specific to the learning task); didactical: pedagogical tools prevent understanding (specific to learners' choice of actions); psycho-genetic: the child's age prevents understanding (specific to the learner's faculties). The notion of "epistemological" obstacle was introduced by

Gaston Bachelard (1938). He identified obstacles as "causes of inertia", causing slowness and disorder. Moreover, Bachelard saw these epistemological obstacles as the driving force behind the evolution of knowledge, since they constitute the rupture that energizes the progress of knowledge. Brousseau also speaks of obstacles, but this time in terms of didactic obstacles: if the pedagogical choices made by the teacher or the educational system are erroneous, they will act as an obstacle to the learning of new knowledge and mislead the pupil. A didactic obstacle is therefore a negative representation of the learning task, induced by previous learning, and hindering new learning. An obstacle thus exists when the "new conceptions" to be appropriated contradict the student's "previous conceptions". "A didactic obstacle is a representation of the task, induced by previous learning, that is the cause of systematic errors and hinders current learning". "An obstacle exists when the new conceptions to be formed contradict the learner's well-established prior conceptions" (Bednarz, Garnier, 1989). We also find this idea of obstacle in Piaget, who saw it from a genetic epistemology point of view: for Piaget, the obstacle is due to psychological limitations.

## Nature of obstacles

Consideration of obstacles from a teaching perspective implies at least partial knowledge of their possible origins. On the one hand, this knowledge enables us to identify these obstacles and, on the other, to make better use of them in teaching. However, knowledge of the origins of these obstacles requires specific skills on the part of the teacher. These can be acquired through training courses at teacher training centers. This is possible if the conditions are right for training that can effectively develop these skills in teachers. It would be illusory to claim to know all the possible origins of obstacles to learning, but it is possible to propose a non-exhaustive list of possible and probable origins of obstacles to learning. In this sense, we can cite the following origins: Language-related obstacles; Obstacles related to the simplification of knowledge for teaching purposes; Obstacles related to the nature of knowledge itself; Obstacles related to notions necessary for understanding the concept under study; Obstacles generated by books and textbooks; Obstacles generated by models proposed by the teacher or reference books; Obstacles related to the learner's learning style; Obstacles related to the educational actions carried out by the teacher.

## Facing the obstacle

The question here is how to deal with the obstacle? Ignore it without ignoring it? Avoid it and get around it by posing the problem differently? Or do we follow a strategy of "dealing with it in order to deal with it"? All these approaches presuppose that the obstacle has been identified, but it should be pointed out that in most cases, the teacher is unaware of the existence of these obstacles, which block his or her students and prevent the acquisition of scientific knowledge. On this subject,

Bachelard said: "I have often been struck by the fact that teachers do not understand that their students do not understand...". Didactic consideration of obstacles: How to deal with obstacles without taking into account students' representations Astolfi (1996) suggests six necessary steps for taking presentations into account:

- Hear them by listening positively to what the students have to say;

- Understand them on the assumption that mistakes are not accidental, but deserve to be analyzed;

- Identify them: Given the unconscious functioning of representations, each person's awareness already contributes to their evolution;

- Comparing them encourages the decentralization of points of view, and reveals to students a diversity of ideas that they may not have imagined in the classroom to explain the same phenomenon;

- Get them talking by provoking socio-cognitive conflicts, which psychology indicates are important levers of intellectual development;

- Monitor their evolution over the short and medium term, throughout compulsory schooling and initially within a single year. It should be noted here that this process of taking representations into account raises certain objections due to the management of didactic time in the face of busy curricula, which requires additional time that the school institution cannot grant.

## The objective-obstacle duality

The concept of objective-obstacle was introduced by JEAN-LOUIS MARTINAND in his doctoral thesis (1982). MARTINAND tried to marry two concepts that are a priori contradictory: obstacles and objectives. Speaking of this coupling, MARTINAND says: "an attempt to bring together two currents, that of pedagogues who seek, through objectives, to make didactic actions more effective, and that of epistemologists who are interested in the difficulties faced by scientific thought".
useful objectives consist in expressing objectives in terms of obstacles that can be overcome, i.e. real difficulties that students encounter and can overcome during the course of the curriculum". Astolfi (1989) has provided us with a process that enables us to implement the objective-obstacle concept:

- Identify obstacles to learning (of which representations are a part), without underestimating or over-valuing them;

- Conversely, and more dynamically, define the intellectual progress corresponding to their eventual crossing;

- Select the obstacle (or obstacles) that you feel you can overcome in the course of a sequence, producing decisive intellectual progress;

- Set the goal of overcoming this obstacle;

- Translate this objective into operational terms using the classic methodologies for formulating objectives;

- Build a system (or several systems) consistent with the objective, as well as remedial procedures in the event of difficulty.

# Didactic situation

## Examples:

- **Race to 20:** each of the opponents must succeed in saying "20" by adding 1 or 2 to the number said by the other; one starts, says 1 or 2 (example: 1), the other continues, adds 1 or 2 to this number (2 for example) and says "3"; in turn the first adds 1 or 2 (1 for example), he says 4, etc. ;

- **The snail race**: On Sunday mornings, a snail climbs a 4-metre high wall. Every day, it climbs 2 metres. Every night, he climbs back down one metre. On which day does he reach the top of the wall?

- **The tangram**: Cut the puzzle as neatly as possible into four pieces. Each student takes possession of one piece. Measure the dimensions of the piece. Enlarge the piece. At the end, you should be able to reconstitute the puzzle with all the enlarged pieces;

- **The side of the puzzle** that measures 4 cm should measure 6 cm when enlarged.

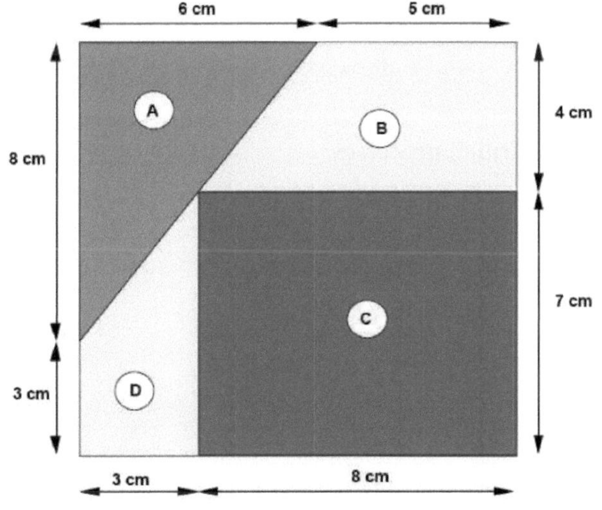


## Definition: (Brousseau)

- **Situation**: A situation is the set of circumstances in which a person finds himself, and the relationships that unite him to his environment;

- **Didactic situation**: Didactic situations are situations that serve to teach. The student's environment is a tool, implemented and manipulated by the teacher;

- **Addidactic situation**: the teacher refuses to intervene as the owner of the knowledge he wants to see appear. The pupil knows that the problem has been chosen to enable him to acquire new knowledge, but he must also know that this knowledge is entirely justified by the internal logic of the situation;

- **Devolution** is the act whereby the teacher makes the student accept responsibility for a learning situation (adidactic) or problem, and accepts the consequences of this transfer himself. It is the process by which the teacher ensures that students assume their share of responsibility for learning.

- **Institutionalization**: the "official" recognition by the student of the object of knowledge, and by the teacher of the student's learning. This is a very important social phenomenon and an essential phase in the didactic process: this double recognition is the object of institutionalization. It's the process in and through which the teacher communicates to students the knowledge or practices they need to retain as the stakes of expected learning.

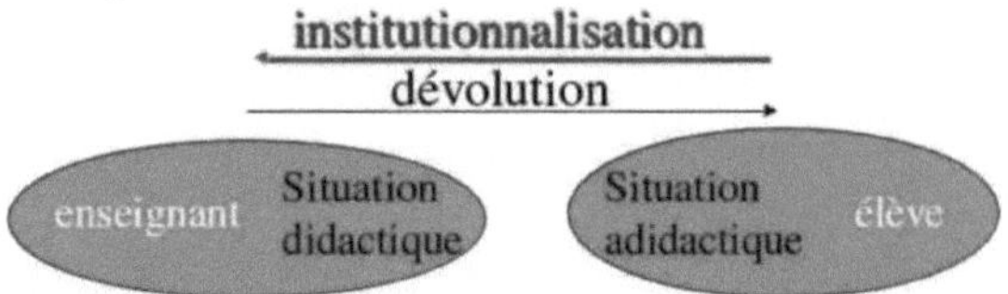


## Analysis of learning processes

There are 4 different phases:

- Action;

- Formulation;

- Validation ;

- Institutionalization .

Knowledge doesn't have the same function during this stage, and students don't have the same relationship to knowledge. Between each phase, there is exchange and regulation between students/groups and the knowledge at stake, and consequently a self-regulated control of learning (dialectic). The 4 phases do not follow each other in regular succession, but are interwoven (back and forth).

## Dialectics of action

The action situation presents the student with a problem whose best solution, under the proposed conditions, is the knowledge to be taught. This allows the student to act on it, and provides him/her with information about his/her action. It's not just a situation of free or imposed manipulation. It enables the student to :

- judge the result of one's action (in-act use of properties) ;
- adjust the action (without teacher intervention);
- learning by adaptation (Piaget);
- establish a (dialectical) dialogue between the child and the situation.

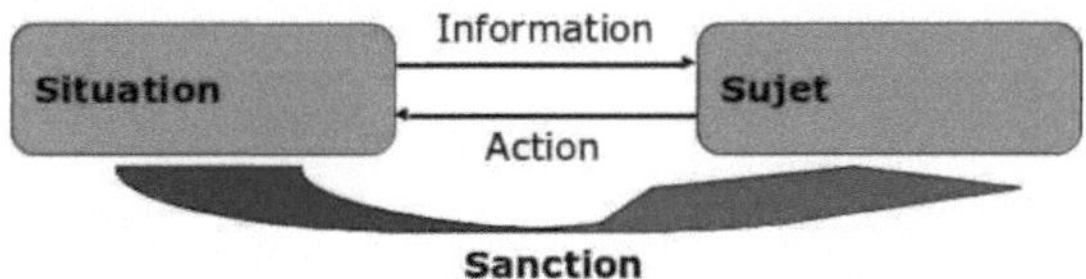


## Dialectics of formulation

The student makes his implicit model explicit, so that this formulation makes sense - to obtain or make obtain a result. It enables :

- Designate, tell, communicate;
- Name the properties ;
- Exchange information (oral or written messages, na "tf or mathematical language) with other students (sender-receivers);
- Create an explicit model ;
- Use signs to formulate common rules, both known and new.

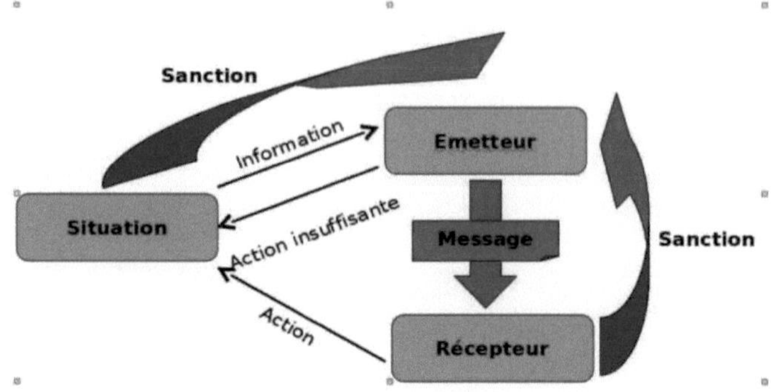

Sanction
Information
Emetteur
Situation
Action insuffisante
Message
Sanction
Action
Récepteur

## Dialectics of validation

- The student must show why the model created is valid: convince (argumentation, demonstration, refutation).
- The student (proposer) submits a mathematical message (model of the situation) as an assertion to an interlocutor (opponent). This is a semantic and syntactic validation.

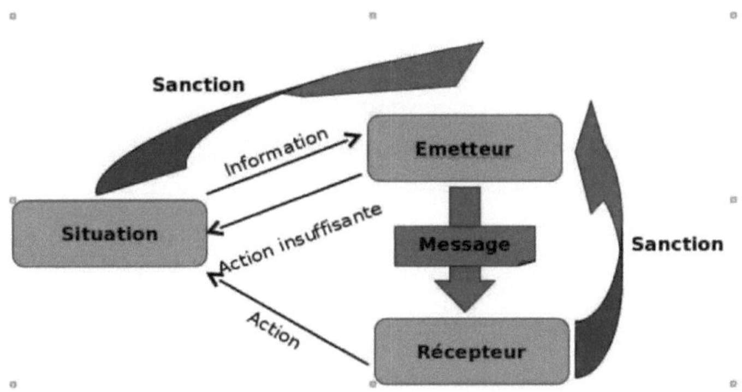


## Dialectics of institutionalization

This is the phase of integrating new knowledge into the mathematical heritage of the class. The teacher conventionally and explicitly sets the cognitive status of the knowledge. Premature institutionalization interrupts the construction of meaning, hinders learning and puts both teacher and students in difficulty Late institutionalization reinforces inaccurate interpretations, slows learning and hinders application.

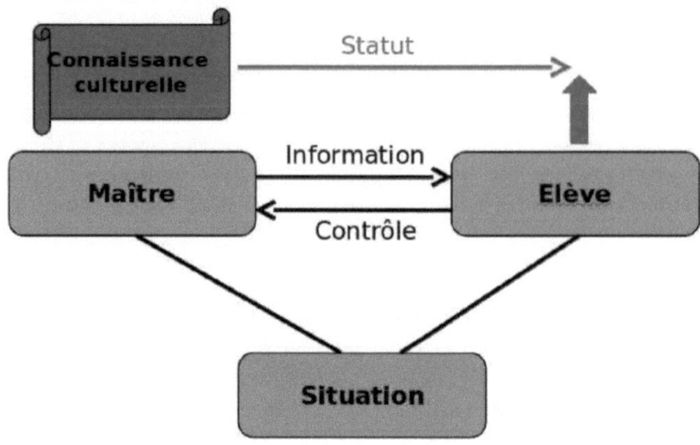


**Validation and assessment :**

- **Validation**: the responsibility of the student, when the situation has been organized for this purpose;
- **Evaluation**: teacher's responsibility, no appeal.

# Chapter 4 : Learning Assessment

# Error analysis

*"Error is not just the effect of ignorance, uncertainty or chance (...) , but the effect of prior knowledge that had its interest, its successes, but which now proves to be false, or simply unsuitable."*

## Introduction

An error is a response that does not conform to what is given as true. The representation of error is first and foremost a question of adequacy to the truth. It's a fairly neutral perception of error. (In. Dictionnaire de pédagogie). In the academic sphere, error is seen as an indicator of whether a learner has objectively acquired a particular skill. In school learning, errors are inevitably present and transitory. Fewer errors are a sign of greater mastery of a given field of knowledge. Given the omnipresence of error in learning, it's essential to analyze its place in modern didactics. Until now, in pedagogy, error has generally been viewed in a negative light. Often likened to a "fault", it had to be punished in order to disappear.

Error is therefore the experience of invalidating initial hypotheses or mental representations. An error occurs because a cognitive process is at work. More precisely, in this process, error marks the phase of destabilization of the initial mental construction, prior to that of reconstruction.

## Error status:

Since the error is indicative of genuine intellectual activity on the part of the student (a strategy of appropriation through the progressive elaboration of representational schemas), it is not blameworthy: it is not the student's fault, it is not a fault. It's not an indication of a lack of knowledge, but of the inadequacy of the student's knowledge to account for reality.

## Theoretical conceptions of error :

- **According to behaviorism**, the aim of teaching is error-free learning. This is achieved through exercises, repetition and reinforcement of "right answers". The student is gradually guided towards achieving a goal (programmed learning);

- **According to constructivism**, learning is a process in which new knowledge can build on or challenge old knowledge. Errors therefore reflect the difficulties that students must overcome in order to produce new knowledge - the so-called cognitive conflict. The correction of an error by a student indicates that he or she has overcome these difficulties by constructing a new response.

## Error types

In fact, there are two types of error:

- **Performance errors**: "silly" mistakes, absent-mindedness or "slips of the tongue", random errors, disruption in the application of a known rule, due to fatigue, stress or emotion caused by the conditions of the assignment. The student knows the rule he should have applied; he is therefore capable of correcting himself. This is what is commonly referred to as a **"fault"**;
- **Competent errors**: revealing the student's intellectual activity ("intelligent errors"): systematic errors that the student is unable to correct, but is able to explain the rule he has applied. With the latter type of deviation from the answer expected by the teacher, error becomes both inevitable (linked to the nature of the student's cognitive development) and useful (it plays a role in the learning process, rather than at the end of it). This is what is commonly referred to as **"error"**.

## Faced with error

Banish, therefore, annotations in the margins of copies: "good", "fair", and other common expressions with moral connotations: "bad" or "good" student, "homework to do". Now there are only mistakes. The didactic value accorded to errors is representative of our conception of learning. This conception of learning rubs off on the pedagogical models we use. For the positive aspects of error to be recognized, the training system must be sensitive to it and consider it as a fundamental element of the school learning process, i.e. it must be "error-tolerant".

# Error status

## Introduction

At school, mistakes are often synonymous with errors. And yet, as we have seen in previous articles, mistakes are a useful part of learning. It is experienced as a fault by both student and teacher. In fact, Astolfi classifies negative reactions to error on the part of the teacher into two categories:

- **Reaction through punishment**: the teacher marks the mistake on the copy, often by underlining it with a red pen. The aim is to draw attention to the mistake and show parents that the teacher has done his job, that he doesn't let mistakes go unnoticed;

- **Punishment by questioning progress**: the teacher feels responsible for the error. For him, if the child has understood, he has done his job well. On the contrary, if the child makes a mistake, it's because he or she has misunderstood, and the fault lies with the teacher.

## Room for error

Instead of punishing or avoiding it, we need to place it at the heart of the teaching process. Mistakes are necessary. It is a stage in the acquisition of knowledge. A pupil can be considered to have made progress if, after making a mistake, he can acknowledge that he has made a mistake, say where and why he went wrong, and how he would do it again without making the same mistakes. To achieve this: the instructive nature of the error, for both teacher and learner, must be clearly explained in the classroom. The teacher must set aside sufficient time for the learner to identify, formulate and explain his or her own mistakes.

## Error handling :

The idea is to work on errors as a tool for educational decision-making.

- **Correcting**: "Correcting is not about judging: it's about helping people learn. It's not about recording and sanctioning deviations from the norm; it's about pointing out specific successes and errors. It's not about performing a terminal act: it's about opening the door to other activities;

- **Noter**: "to appreciate by a numerical mark" (definition given by the Petit Robert). Several notation tools can be used: numerical grade, letter grade, color grade.

- **Annotate**: Add critical or explanatory notes.

- **Evaluation**: In the school context, it means confronting a student's production with a set of criteria that are predefined, objective (eliminating moral judgment, but not judgment) and explicit (known).

## Who corrects, what and why?

- **Who:** The teacher and the learner who produced the work, possibly another student or a group of students. An explicit contract must define each person's task (which may vary according to the type of production). Correction by the teacher alone is of little benefit to the learner. The only useful correction is the one made by the learner. Training students to reread their work during class or at the end of a lesson encourages them to take their work as the object of study, and to rectify it if necessary. The teacher checks the learner's correction.

- **What:** All work must be checked and all written work must be corrected, whether it's written work, grammar exercises or vocabulary. Whatever the medium: copies, notebooks.

- **Why**

  ○ *For the teacher*: it allows you to check expected results, verify the acquisition of skills, analyze and correct errors.

  ○ *For the learner:* it enables progress towards target skills by reinvesting knowledge.


## Error types

Jean-Pierre Astolfi has identified 7 types of errors made by students:

- **Understanding instructions**: The child does not understand the instruction and cannot fulfill the didactic contract. The problem may stem from the difficulty of the statement;

- **School habits and poor decoding**: Students operate on the principle of mechanics (customary didactics), making it difficult for them to respond to instructions that are out of the ordinary.

- **Cognitive overload**: This is a memory problem that most often arises when the child has to process several pieces of information at the same time. From one subject to another: the student must systematically reinvest his or her knowledge in each subject. This requires incessant gymnastics on the part of the brain, sometimes leading to errors.

- **Errors reflecting alternative conceptions:** Students don't wait for a leçon to explain themselves in relation to a given problem. They already have intellectual representations of the various notions they are going to study.

- **Errors related to intellectual operations:** The child does not yet have the necessary skills (reversibility, transitivity, working with states or transformations) to respond to the teacher's requests.

- **Errors involving procedures:** the student uses procedures not necessarily expected by the teacher. But often the approach is perceived as an error by the teacher, who expects a precise answer.

  **Errors caused by the complexity of the content:** The teacher needs to explain instructions to the pupils. If the child doesn't understand the instructions or the context, there will be an error.

## The student and error

We've seen that students feel differently about mistakes. This may be due to the influence of their environment.

- *Class influence :*
  - In the classroom, the student exists as an individual, but also as part of the class. They are part of a whole.
  - The other students in the class exert involuntary pressure.
  - The child is always working under the influence.
- *Teacher influence :*
  - Some children are also under pressure from their teachers.
  - In order to fulfill his role as a student and accept the didactic contract, the child must answer the questions posed.
  - The teacher must allow the child to express himself freely, and ensure that every word, good or bad, is respected.

We've seen how the teacher perceives the error, but it's above all necessary to see how the pupil reacts to his mistakes. The child may feel either a sense of surpassing or a sense of being surpassed:

- *An overachiever:*
  - Fear of taking the risk of making mistakes.
  - Mistakes are perceived as failures: mistakes as faults.
  - Errors can be seen as a barrier or obstacle to children's learning.
  - Fear of error leads to a lack of participation.
  - Mistakes are even associated with anxiety for some children (= allergies according to J.P. Astolfi).
  - These reactions can be explained by the child's character (shy, introverted).
- *Overcoming obstacles:* According to G. Bachelard, obstacles help us to construct rational thought. For him:
  - "The mind can only be formed by reforming itself";
  - A student who makes mistakes is not necessarily a bad student;
  - Parental education will also influence the child's choices in the face of error. It's by making mistakes that children learn to grow;
  - Error = challenge and positive questioning;

    Error is not always a valid evaluation criterion

# Assessing learning

## The principle of evaluation :

Assessment is at the heart of training. There can be no valid teaching/learning strategy without the implementation of a coherent assessment system. The organization of this system is only possible if pedagogical intentions have been clearly defined in the form of teaching and learning objectives. Assessment is at the service of a pedagogy of success. Its aim is not to select the best, but to help as many people as possible to achieve their objectives.

## Types of assessment :

- *Diagnostic*: Situate the learner at the start of a sequence (initial test). The aim is to check that learners have acquired the necessary skills to follow the sequence (prerequisites);

- *Formative*: to check the level of acquisition during or after the sequence. In this case, it's a tool for diagnosing difficulties and successes. The aim is to facilitate learning. During these assessments, which must be frequent, the student has the right to make mistakes. Errors and blockages are exploited by the teacher to re- - explain. It's a privileged moment for dialogue that should enable: the student to know where he stands; the teacher to propose: activities to help students in difficulty, more complex activities for high-achieving students;

- *Summative*: When the teacher considers that students have had sufficient practice, he or she proposes an assessment in which the learner must demonstrate that he or she has achieved the objective. In this case, there is no room for error. Assessment results in a grade, or recognition of prior learning (when the decision is made to move the student on to a higher class. . .).

## Evaluation criteria.

The basic principle Assessment focuses on: skills, knowledge and know-how, attitudes. Assessment of student work must not be subjective. The teacher must be able to justify the assessment or mark awarded. It is therefore important to specify the assessment criteria that define the student's work contract. The student must be informed of these criteria or indicators of success from the outset. In the course of training, it will be important, if not essential, to refine the criteria according to the prerequisites and achievements of the students, and to eliminate any ambiguity in the formulation of these criteria.

At the start of learning. The teacher: involves his students in his teaching project;

gives meaning to the activities he will have to propose throughout the didactic project. These are teaching aids. The learner: will know what is expected of him/her, and will prepare accordingly. The grids will be a facilitator and learning aid.

At the end of the learning process: For the teacher, these will be tools for criterion-based assessment of the whole class and of each individual student in relation to the final competency, but also in relation to each intermediate competency. In this way, the teacher will know the extent to which his class and each of his pupils have reinvested the content he has taught.

These grids will give the teacher a clear view of: the difficulties (obstacles) encountered. the performances achieved and the results obtained by the class and by each pupil. If the objective has been reached (mastery of the skill) : Move on to the next didactic unit. If objective not reached: construction of remediation and regulation sequences. For the learner, these are "contract sheets" or grids against which he or she can self-assess. These reference sheets can be used to correct erroneous statements in the learner's work.

## Subjective assessment

Grading an assignment is a difficult art, and one that can leave a great deal of room for subjectivity. If papers are graded in a certain order by one teacher and in the opposite order by another, the graders tend to over-rate the papers graded first and under-rate the papers graded last.

There is also an assimilation effect, which consists in comparing a mark with a previous mark. For example, teachers are asked to assess six papers of roughly equivalent level. But on each of them appears a mark, supposedly obtained by the same pupil some time previously. If this previous mark is high, the mark is on average 2 points higher than if the previous copy is low (averages of 11.86 and 9.84). Similarly, when confronted with a series of identical papers, teachers evaluate them differently depending on whether they come from a strong or weak class. The same effect is seen depending on the high school from which the homework originated.

Finally, the contrast effect is an inverse process. A copy is generally over-rated when corrected after a weak copy, and under-rated when corrected after a strong copy. Grading is always random. Different examiners (correctors) do not evaluate the same objects in the same faj con way, do not use grade scales in the same way, and produce judgments that are not stable over time. Rather than succumb to the myth of the "true" grade, *isn*'t it better to focus on what assessment is (or should be!) used for when integrated into the training process? Formative assessment is assessment that plays a regulatory role in teaching and learning. Through the information it provides, assessment helps to regulate the didactic process (teaching and/or learning).

If it's a learning process, the aim is to guide the student, enabling him to recognize, understand and correct his own mistakes. (Corrective function), informing him/her of the stages reached or not, while at the same time informing the teacher of the real effects of his/her pedagogical actions (Regulatory function).

The aim of assessment is not so much to describe reality as it is, but rather to help people become what they could be (Transformative aim). Indeed, if the essential aim is to help students identify, analyze and understand their errors, so as to avoid making them in the future, evaluation must have a theoretical model that makes this analysis possible and, more generally, a model of the student's cognitive functioning.

## Effective evaluation.

An effective evaluation is one that achieves its goal. Good evaluation means first and foremost understanding what you're doing, and why you're doing it. The difficulty lies in knowing from which point of view to judge the relevance of a practice: there is no such thing as a correct evaluation in itself. But there are relevant evaluations, according to a given intention and for a specific use.

In formative evaluation, teachers' task is to contribute to the positive development of students by facilitating their learning. It is this pedagogical objective that gives formative evaluation its meaning.

And an effective evaluation is one that sheds light, hence the need to spell out expectations in terms of

skills and know-how, analyze and interpret errors, identify the characteristics of the participants, and make a precise diagnosis of each person's achievements and shortcomings, strengths and weaknesses.

Grading seems unreliable. Today, the imperative is to assess as objectively as possible the extent to which learners have achieved the pedagogical objectives assigned to them. Assessment must not stop at quantitative statements of our students' results (grades, averages). It must be able to give a rigorous interpretation of these results, those that really deserve to be taken into consideration.

In fact, to evaluate, you need to observe as rigorously as possible, and interpret as pertinently as possible.

## Evaluation is information.

What's more, the teacher's assessment is not simply a matter of measuring, judging and deciding, but also of providing useful information to the student to facilitate learning and help him or her become aware of errors, by providing benchmarks for correct self-evaluation. Under these conditions, evaluation is descriptive. As such, it is uniquely compatible with the helping relationship advocated by the learner-centered pedagogical approach. The aim is to inform in order to help, rather than to judge in order to decide. To sum up, an effective evaluation has the threefold characteristic of being :

• *Comprehension:* (able to interpret the "measured" situation) ;

• *Conscientious* (providing enlightening reference points for the student, rather than just chastising him or her);

• *Trainer* (concerned with providing the tools for success).

# References


Astolfi, A. (1996). Discontinuous control of nonholonomic systems. Systems & control letters, 27(1), 37-45.

Audigier, F. (1988). Didactique de l'histoire, de la géographie et des sciences sociales: propos introductifs. Revue française de pédagogie, 5-9.

Bachelard, G. (1938). La formation de l'esprit scientifique, Paris. J. Vrin.

Baruk, S. (2016). The age of the captain. On error in mathematics. Media Diffusion.

Brousseau, G. (1990). Le contrat didactique: le milieu. Recherches en didactique des mathématiques, 9(9.3), 309-336.

Chevallard, Y., & Johsua, M. A. (1985). La transposition didactique: du savoir savant au savoir enseigné. La Pensée Sauvage.

Clerc, J. B., Minder, P., & Roduit, G. (2006). La transposition didactique. Dictionnaire Des Concepts Fondamentaux Des Didactiques.

Develay, M. (1992). De l'apprentissage à l'enseignement: pour une épistémologie scolaire. Éditions ESF,.

Halté, J. F. (1992). La didactique du français.

Giordan, A. (1996). ¿ Cómo ir más allá de los modelos constructivistas? La utilización didáctica de las concepciones de los estudiantes. Revista Investigación en la Escuela, 28, 7-22.

Giordan, A., & De Vecchi, G. (1987). Les origines du savoir. From learners' conceptions to scientific concepts. Neuchâtel-Paris: Delachaux et Nestlé.

Le Pellec, J., & Marcos-Alvarez, V. (1991). Enseigner l'histoire: un métier qui s' apprend. FeniXX.

Martinand, J. L. (1985). Sur la caractérisation des objectifs de l'initiation aux sciences physiques. Aster: Recherches en didactique des sciences expérimentales, 1(1), 141-154.

Reuter, Y., & Lahanier-Reuter, D. (2007). Discipline analysis: some problems for didactic research. La didactique du français. Les voies actuelles de la recherche, 27-42.

Tardy, M. (1993). "La transposition didactique". In J. Houssaye, (Ed.), La pédagogie : une encyclopédie pour aujourd'hui (pp. 51-60). Paris: E.S.F.